by Mona Lee and Kathryn Corbett

illustrated by Laura Freeman

Orlando Boston Dallas Chicago San Diego

www.harcourtschool.com

Bake sale, bake sale,
bake sale today!
Cookies cost $1.00.
How will you pay?

Brown cows, blue birds,
tan dogs, too.
White cats are my favorite.
How about you?

Li wants the brown cow
from the cookie tray.
"$1.00, please," says Tasha.
Li knows how to pay.

"100 pennies equal
$1.00–you are right!"
Li picks up the brown cow
and takes a big bite.

Nick wants the blue bird
from the cookie tray.
“$1.00, please,” says Tasha.
Nick knows how to pay.

“20 nickels equal
$1.00–you are right!”
Nick picks up the blue bird
and takes a big bite.

Meg wants the white cat
from the cookie tray.
"$1.00, please," says Tasha.
Meg knows how to pay.

"10 dimes equal
$1.00—you are right!"
Meg picks up the white cat
and takes a big bite.

Kurt wants the tan dog
from the cookie tray.
"$1.00, please," says Tasha.
Kurt knows how to pay.

"4 quarters equal
$1.00—you are right!"
Kurt picks up the tan dog
and takes a big bite.

Tasha spends her money
to get her pets a treat.
Her cow, and bird, and cat, and
dog all just love to eat.